Times Tables Tricks and Tips

For Kids and Adults

FM Bill Jordan

Times Tables Tricks and Tips

Times Tables Tricks and Tips : For Kids and Adults
First edition 2019. 2nd edition 2021.
Copyright © 2021 Bill Jordan.
Written by Bill Jordan.

Table of Contents

Introduction..7
    Why Learn the Times Tables?..7
    Who is this book for?..7
    Better Ways of Learning...8
    Ways of Learning Tables...8
The Grid..9
Odds And Evens...10
The Plus Tables..11
The Plus Table..15
    Adding Digits..16
    Subtracting Numbers..17
Multiplication...18
    What is multiplication?...18
    How many Times Tables are there?.....................................19
Multiplying by Zero..20
Multiplying by One...20
Multiplying by Ten..21
How many items to learn?..21
2 Times Tables..23
    Halving Even Numbers..24
    Doubling Numbers...25
3 Times Tables..25
    3 x Table..26
4 Times Tables..27
    4 Times Tables and Other Tables...28
5 Times Tables..28
6 Times Tables..30
    6 Times Tables and Even Numbers......................................31
    6 Times Tables and Other Tables...32
7 Times Tables..32
    Tip..33
8 Times Tables..33
    8 Times Tables and Other Tables...34
    Tip..34
9 Times Tables..34
    9 Times Tables and Other Tables...36

Times Tables Tricks and Tips

- 10 Times Tables...36
  - Multiplication Patterns.....................................37
- 11 Times Tables...38
- 12 Times Tables...39
  - 12 Times Tables and Other Tables.......................40
  - Multiples of 12..41
- Cycles..41
  - Prime Numbers..43
- Squares..43
- Division...51
  - What is Division?...51
  - Division with a Remainder.................................52
  - Division by 10..53
  - Division by 5...53
- Square Roots..53
- More Plus Tables..54
  - The Minus Tables...56
- More Minus Tables...60
- Summary..64
  - Where to go from here?...65
- About the Author..66

Introduction

Learning the Times Tables can be challenging.

Why Learn the Times Tables?

Understanding the times tables helps in many jobs.

For example,

- Accountant
- Architect
- Graphics designer
- Computer programmer
- Engineer
- Many others

The times table are a building block for learning other branches of mathematics.

Learning the times tables can help develop your memory.

Who is this book for?

- Parents who have children learning the Times Tables.
- Children who are learning the Times Tables.
- Older children who want to sharpen up their Times Tables.
- Adults who want to sharpen up their knowledge and understanding of the Times Tables.

Times Tables Tricks and Tips

Better Ways of Learning

I have memories of being in a classroom with other kids chanting two twos are four, two threes are six etc.

I don't think this was the only or best way of learning the tables.

Learning the tables is a **memory** game. It is much easier to remember something if it has **meaning** to you.

Rhyme can create meaning.

For example,
Thirty days hath September, April, June and November

This is easier to learn because September and November rhyme.

Visual patterns are easier to remember. There is a saying that a picture says a thousand words.

If you **understand** how tables work, there will be:

- Less to learn.
- Learning them will be easier.

Ways of Learning Tables

- Learn them with rhymes.
- Learn then visually.
- Understand how they work.
- Recognising patterns.

Rhymes are Ok. However this book will concentrate on the other methods.

If you have any suggested corrections you can email Bill Jordan at **swneerava@gmail.com**.

1	2	3	4	5	6	7	8	9	10
11	12	13	14	15	16	17	18	19	20
21	22	23	24	25	26	27	28	29	30
31	32	33	34	35	36	37	38	39	40
41	42	43	44	45	46	47	48	49	50
51	52	53	54	55	56	57	58	59	60
61	62	63	64	65	66	67	68	69	70
71	72	73	74	75	76	77	78	79	80
81	82	83	84	85	86	87	88	89	90
91	92	93	94	95	96	97	98	99	100

The Grid

Here is our basic 10 x 10 grid. It has ten **rows** and ten **columns**. It is filled with the numbers from 1 to 100.

On the grid we will be able to recognise many patterns.

It is traditional for the Times Tables to go up to x 12. Why does it go to 12 rather than 10, which is a nice round number?

In the Imperial system there are 12 inches in a foot.

When the metric system is used, its not so important to use to 11 and 12 Times Tables. The 11 x and 12 x tables will be covered later.

Start at a number in the middle of the grid, for example 45.

Moving right will add one to the number.
Moving left will subtract one from the number.

Moving down will add 10 to the number.
Moving north will subtract ten from the number.

Times Tables Tricks and Tips

Moving down and left will add 9 to the number.
Moving down and right will add 11 to the number.

Moving up and left will subtract 11 from the number.
Moving up and right will subtract 9 from the number.

Odds And Evens

Starting with the number 1, every second number is odd. Starting with the number 2, every second number is even.

1	2	3	4	5	6	7	8	9	10
11	12	13	14	15	16	17	18	19	20
21	22	23	24	25	26	27	28	29	30
31	32	33	34	35	36	37	38	39	40
41	42	43	44	45	46	47	48	49	50
51	52	53	54	55	56	57	58	59	60
61	62	63	64	65	66	67	68	69	70
71	72	73	74	75	76	77	78	79	80
81	82	83	84	85	86	87	88	89	90
91	92	93	94	95	96	97	98	99	100

- If you add 2 **even** numbers you always get an even number.

  For example, 2 + 4 = 6.

- If you add 2 **odd** numbers you always get an even number.
  For example, 3 + 5 = 8.

- If you add an **even** number and an **odd** number you always get an odd number.
  For example, 2 + 3 = 5.

The Plus Tables

You have heard of the Times Tables, but have you heard of the plus tables?

We notice that some of these sums are the same.

For example, **4 + 3 = 7** is the same as **3 + 4 = 7**.

This is because the order of the 4 and 3 doesn't matter. The fancy way of saying this is to say that addition is **commutative**. We can use this information to make the plus tables smaller.

These tables are designed to give you feel for the relative **sizes** of different numbers.

2 Plus Table		
1+1=2	1	1

These are numbers that add up to 2.

3 Plus Table			
	1	2	3
1+2=3	1		2
2+1=3		2	1

These are numbers that add up to 3.

Times Tables Tricks and Tips

4 Plus Table				
	1	2	3	4
1+3=4	1		3	
2+2=4	2		2	
3+1=4	3			1

These are numbers that add up to 4.

5 Plus Table					
	1	2	3	4	5
1+4=5	1		4		
2+3=5	2		3		
3+2=5	3			2	
4+1=5	4				1

These are numbers that add up to 5.

6 Plus Table						
	1	2	3	4	5	6
1+5=6	1			5		
2+4=6	2			4		
3+3=6	3			3		
4+2=6	4				2	
5+1=6	5					1

These are numbers that add up to 6.

7 Plus Table

	1	2	3	4	5	6	7
1+6=7	1			6			
2+5=7	2				5		
3+4=7	3				4		
4+3=7	4				3		
5+2=7	5					2	
6+1=7	6						1

These are numbers that add up to 7.

8 Plus Table

	1	2	3	4	5	6	7	8
1+7=8	1			7				
2+6=8	2				6			
3+5=8	3				5			
4+4=8	4				4			
5+3=8	5					3		
6+2=8	6						2	
7+1=8	7							1

These are numbers that add up to 8.

Times Tables Tricks and Tips

9 Plus Table									
	1	2	3	4	5	6	7	8	9
1+8=9	1				8				
2+7=9		2				7			
3+6=9			3				6		
4+5=9				4			5		
5+4=9				5			4		
6+3=9					6			3	
7+2=9					7				2
8+1=9						8			1

These are numbers that add up to 9.

10 Plus Table										
	1	2	3	4	5	6	7	8	9	10
1+9=10	1				9					
2+8=10		2				8				
3+7=10			3				7			
4+6=10				4			6			
5+5=10				5			5			
6+4=10					6			4		
7+3=10					7			3		
8+2=10					8				2	
9+1=10					9				1	

These are numbers that add up to 10.

One of the most important plus tables is the 10 + table.

	1	2	3	4	5	6	7	8	9	10
1+	2	3	4	5	6	7	8	9	10	11
2+	3	4	5	6	7	8	9	10	11	12
3+	4	5	6	7	8	9	10	11	12	13
4+	5	6	7	8	9	10	11	12	13	14
5+	6	7	8	9	10	11	12	13	14	15
6+	7	8	9	10	11	12	13	14	15	16
7+	8	9	10	11	12	13	14	15	16	17
8+	9	10	11	12	13	14	15	16	17	18
9+	10	11	12	13	14	15	16	17	18	19
10+	11	12	13	14	15	16	17	18	19	20

The Plus Table

You can add 2 times and see the result.

Step 1: Choose a number in the left most column.
Step 2: Choose a number in the top most row.
Step 3: Look along the row of the first number.
Step 4: Look along the column of the second number.
Step 5: Where they match is the answer!

Times Tables Tricks and Tips

	1	2	3	4	5	6	7	8	9	10
1+	2	3	4	5	6	7	8	9	10	11
2+	3	4	5	6	7	8	9	10	11	12
3+	4	5	6	7	8	9	10	11	12	13
4+	5	6	7	8	9	10	11	12	13	14
5+	6	7	8	9	10	11	12	13	14	15
6+	7	8	9	10	11	12	13	14	15	16
7+	8	9	10	11	12	13	14	15	16	17
8+	9	10	11	12	13	14	15	16	17	18
9+	10	11	12	13	14	15	16	17	18	19
10+	11	12	13	14	15	16	17	18	19	20

For example,
6 + 5 = ?
Look up 6 in the left most column.
Look up 5 in the top most row.
They match at 11!

Adding Digits

It is easier to add up two digits if they add up to less than ten.

This is true if numbers have more than one digit, provided all matching numbers add up to less than ten.

For example:

12345678
+
87654321
=
99999999

Subtracting Numbers

It is easier to subtract a smaller digit from a larger one if they add up to less than ten.

This is true if numbers have more than one digit, provided that is true for all matching digits.

For example:

864
-
123
=
741

Times Tables Tricks and Tips

Multiplication

What is multiplication?

Multiplication is really a form of addition. Addition is repeated a certain number of times.

For example,
10 + 10 + 10 is the same as **10 x 3**.

	1	2	3	4	5	6	7	8	9	10	11	12
1x	1	2	3	4	5	6	7	8	9	10	11	12
2x	2	4	6	8	10	12	14	16	18	20	22	24
3x	3	6	9	12	15	18	21	24	27	30	33	36
4x	4	4	8	16	20	24	28	32	36	40	44	48
5x	5	10	15	20	25	30	35	40	45	50	55	60
6x	6	12	18	24	30	36	42	48	54	60	66	72
7x	7	14	21	28	35	42	49	56	63	70	77	84
8x	8	16	24	32	40	48	56	64	72	80	88	96
9x	9	18	27	36	45	54	63	72	81	90	99	108
10x	10	20	30	40	50	60	70	80	90	100	110	120
11x	11	22	33	44	55	66	77	88	99	110	121	132
12x	12	24	36	48	60	72	84	96	108	120	132	144

This is the **Times Tables** up to times 12.

This is a *lookup* table.

You can multiply 2 times and see the result.

Step 1: Choose a number in the left most column.
Step 2: Choose a number in the top most row.
Step 3: Look along the row of the first number.
Step 4: Look along the column of the second number.

Step 5: Where they match is the answer!

	1	2	3	4	5	6	7	8	9	10	11	12
1x	1	2	3	4	5	6	7	8	9	10	11	12
2x	2	4	6	8	10	12	14	16	18	20	22	24
3x	3	6	9	12	15	18	21	24	27	30	33	36
4x	4	4	8	16	20	24	28	32	36	40	44	48
5x	5	10	15	20	25	30	35	40	45	50	55	60
6x	6	12	18	24	30	36	42	48	54	60	66	72
7x	7	14	21	28	35	42	49	56	63	70	77	84
8x	8	16	24	32	40	48	56	64	72	80	88	96
9x	9	18	27	36	45	54	63	72	81	90	99	108
10x	10	20	30	40	50	60	70	80	90	100	110	120
11x	11	22	33	44	55	66	77	88	99	110	121	132
12x	12	24	36	48	60	72	84	96	108	120	132	144

For example,
7 x 8 = ?
Look up 7 in the left most column.
Look up 8 in the top most row.
They match at 56!

How many Times Tables are there?

We can go up to 10 x 10 (and leave 11x and 12x tables till later). It looks like to learn the tables, we would need to know the following.

Times Tables Tricks and Tips

1x1	1x2	1x3	1x4	1x5	1x6	1x7	1x8	1x9	1x10
2x1	2x2	2x3	2x4	2x5	2x6	2x7	2x8	2x9	2x10
3x1	3x2	3x3	3x4	3x5	3x6	3x7	3x8	3x9	3x10
4x1	4x2	4x3	4x4	4x5	4x6	4x7	4x8	4x9	4x10
5x1	5x2	5x3	5x4	5x5	5x6	5x7	5x8	5x9	5x10
6x1	6x2	6x3	6x4	6x5	6x6	6x7	6x8	6x9	6x10
7x1	7x2	7x3	7x4	7x5	7x6	7x7	7x8	7x9	7x10
8x1	8x2	8x3	8x4	8x5	8x6	8x7	8x8	8x9	8x10
9x1	9x2	9x3	9x4	9x5	9x6	9x7	9x8	9x9	9x10
10x1	10x2	10x3	10x4	10x5	10x6	10x7	10x8	10x9	10x10

This is a total of 100 items. However, there are some shortcuts which will make it **much** less.

Multiplying by Zero

You do not need the learn the Zero Times Tables. You do need to remember that a number times zero is always zero.

0 x 7 is 0.
0 x 27 is 0.
0 x 1,000,000 is 0.
Something x 0 = **0**.

The Zero Times Tables are not tables at all, we do not need to learn them.

Multiplying by One

You do not need the learn the 1 Times Tables. You do need to remember that a number times 1 is always the same as itself.

1 x 7 is 7.
1 x 27 is 27.
1,000,000 x 1 is 1,000,000.
Something x 1 = **Something**.

The 1 Times Tables are not really tables at all, we do not need to learn them.

Multiplying by Ten

10 Times Tables are very easy. All we have to do to multiply by 10 is to place a zero at the end of the number.

For example, if we want to multiply 4 x 10, we put a zero after the 4 to get 40.

How many items to learn?

Without the **x1** row and the **x10** row that leaves:

2x1	2x2	2x3	2x4	2x5	2x6	2x7	2x8	2x9	2x10
3x1	3x2	3x3	3x4	3x5	3x6	3x7	3x8	3x9	3x10
4x1	4x2	4x3	4x4	4x5	4x6	4x7	4x8	4x9	4x10
5x1	5x2	5x3	5x4	5x5	5x6	5x7	5x8	5x9	5x10
6x1	6x2	6x3	6x4	6x5	6x6	6x7	6x8	6x9	6x10
7x1	7x2	7x3	7x4	7x5	7x6	7x7	7x8	7x9	7x10
8x1	8x2	8x3	8x4	8x5	8x6	8x7	8x8	8x9	8x10
9x1	9x2	9x3	9x4	9x5	9x6	9x7	9x8	9x9	9x10

Without the **x1** column and the **x10** column that leaves:

Times Tables Tricks and Tips

2x2	2x3	2x4	2x5	2x6	2x7	2x8	2x9
3x2	3x3	3x4	3x5	3x6	3x7	3x8	3x9
4x2	4x3	4x4	4x5	4x6	4x7	4x8	4x9
5x2	5x3	5x4	5x5	5x6	5x7	5x8	5x9
6x2	6x3	6x4	6x5	6x6	6x7	6x8	6x9
7x2	7x3	7x4	7x5	7x6	7x7	7x8	7x9
8x2	8x3	8x4	8x5	8x6	8x7	8x8	8x9
9x2	9x3	9x4	9x5	9x6	9x7	9x8	9x9

Now there are only 64 things to learn. Our table is getting smaller all the time!

The order in which the numbers are makes no difference.

For example 2x3 is the same as 3x2.

We can use this information to make the Times Tables smaller.

2x2	2x3	2x4	2x5	2x6	2x7	2x8	2x9
	3x3	3x4	3x5	3x6	3x7	3x8	3x9
		4x4	4x5	4x6	4x7	4x8	4x9
			5x5	5x6	5x7	5x8	5x9
				6x6	6x7	6x8	6x9
					7x7	7x8	7x9
						8x8	8x9
							9x9

Now there are only 36 things to learn!

2x2							
3x2	3x3						
4x2	4x3	4x4					
5x2	5x3	5x4	5x5				
6x2	6x3	6x4	6x5	6x6			
7x2	7x3	7x4	7x5	7x6	7x7		
8x2	8x3	8x4	8x5	8x6	8x7	8x8	
9x2	9x3	9x4	9x5	9x6	9x7	9x8	9x9

This is another of representing the same table. In each case there are only 36 equations in the tables.

When the 12x tables are added there are another 10 things to learn. You don't need to memorise 12x1 and 12x10.

When the 11x tables are added, you only need learn 11x11.

So there are only **47** items in the Times Tables up to 12 you need to learn.

2 Times Tables

	1	2	3	4	5	6	7	8	9	10	11	12
2x	2	4	6	8	10	12	14	16	18	20	22	24

Multiples of 2 are all even numbers.

The last digit is always 0, 2, 4, 6 or 8.

Times Tables Tricks and Tips

2 x Table									
1	**2**	3	**4**	5	**6**	7	**8**	9	**10**
11	**12**	13	**14**	15	**16**	17	**18**	19	**20**
21	**22**	23	**24**	25	26	27	28	29	30
31	32	33	34	35	36	37	38	39	40
41	42	43	44	45	46	47	48	49	50
51	52	53	54	55	56	57	58	59	60
61	62	63	64	65	66	67	68	69	70
71	72	73	74	75	76	77	78	79	80
81	82	83	84	85	86	87	88	89	90
91	92	93	94	95	96	97	98	99	100

Multiples of 2 appear as 5 straight columns in the grid.

This is a visual pattern that is **easy** to remember.

Halving Even Numbers

If you know the 2 Times Tables you can **halve** even numbers up to 24.

12/2 = 6
is the mathematical way of saying **half** of 12 is 6.

Here are half of even numbers up to 12.

Halving Table												
Half of 2 is 1	1	2									
Half of 4 is 2	1	2	3	4							
Half of 6 is 3	1	2	3	4	5	6					
Half of 8 is 4	1	2	3	4	5	6	7	8			
Half of 10 is 5	1	2	3	4	5	6	7	8	9	10	
Half of 12 is 6	1	2	3	4	5	6	7	8	9	10	11	12

Doubling Numbers

Doubling a number is the same as multiplying it by 2.

For example,

Double 4 is 2x4 is 8.

3 Times Tables

	1	2	3	4	5	6	7	8	9	10	11	12
3x	3	6	9	12	15	18	21	24	27	30	33	36

Starting with the first, every second multiple of 3 is odd.

Starting with the second, every second multiple of 3 is even.

Adding the digits together creates a pattern: 3,6,9,3,6,9,3,6,9 etc.

3 x Table

3 x Table									
1	2	**3**	4	5	**6**	7	8	**9**	10
11	**12**	13	14	**15**	16	17	**18**	19	20
21	22	23	**24**	25	26	**27**	28	29	**30**
31	32	**33**	34	35	**36**	37	38	**39**	40
41	42	**43**	44	**45**	46	47	**48**	49	50
51	52	53	**54**	55	56	**57**	58	59	**60**
61	62	**63**	64	65	**66**	67	68	**69**	70
71	**72**	73	74	**75**	76	77	**78**	79	80
81	82	83	**84**	85	86	**87**	88	89	**90**
91	92	**93**	94	95	**96**	97	98	**99**	100

Multiples of 3 appear as diagonal lines in the grid.

This is a visual pattern that is **easy** to remember.

3 x Table			
3x1=3	1	2	**3**
3x2=6	4	5	**6**
3x3=9	7	8	**9**
3x4=12	10	11	**12**
3x5=15	13	14	**15**
3x6=18	16	17	**18**
3x7=21	19	20	**21**
3x8=24	22	23	**24**
3x9=27	25	26	**27**
3x10=30	28	29	**30**
3x11=33	31	32	**33**
3x12=36	34	35	**36**

4 Times Tables

	1	2	3	4	5	6	7	8	9	10	11	12
4x	4	8	12	16	20	24	28	32	36	40	44	48

Multiples of 4 are always even numbers.

Times Tables Tricks and Tips

4 x Table									
1	2	3	**4**	5	6	7	**8**	9	10
11	**12**	13	14	15	**16**	17	18	19	**20**
21	22	23	**24**	25	26	27	**28**	29	30
31	**32**	33	34	35	**36**	37	38	39	**40**
41	42	43	**44**	45	46	47	**48**	49	50
51	52	53	54	55	56	57	58	59	60
61	62	63	64	65	66	67	68	69	70
71	72	73	74	75	76	77	78	79	80
81	82	83	84	85	86	87	88	89	90
91	92	93	94	95	96	97	98	99	100

We can see a pattern.
The last digit ends in 4 or 8 on the odd rows.
The last digit ends in 2, 6 or 0 on the even rows.

4 Times Tables and Other Tables

	1	2	3	4	5	6	7	8	9	10	11	12
2x	2	4	6	8	10	12	14	16	18	20	22	24
4x	4	8	12	16	20	24	28	32	36	40	44	48

The 4x table is twice the 2x table.

5 Times Tables

	1	2	3	4	5	6	7	8	9	10	11	12
5x	5	10	15	20	25	30	35	40	45	50	55	60

When you multiply 5 by an odd number, the answer will end in 5.

If you multiply 5 by an even number, the answer will end in zero.

There is an easy way to multiply by 5.

- Step 1: Multiply by 10 (very easy, you put a zero at the end).
- Step 2: Halve this number.

If you are multiplying 5 by an even number, halving the number is simple.

- Half of 2 is 1.
- Half of 4 is 2.
- Half of 6 is 3.
- Half of 8 is 4.

For example, **8x5 = ?**.

- Step 1: 8x10 = 80
- Step 2: Half of 80 is 40.

If you are multiplying 5 by an odd number, halving the number is a little more complex.

- Half of 10 is 5.
- Half of 30 is 15.
- Half of 50 is 25.
- Half of 70 is 35.
- Half of 90 is 45.

For example, **7x5 = ?**

- Step 1: 7x10 = 70
- Step 2: Half of 70 is 35.

Times Tables Tricks and Tips

5 x Table									
1	2	3	4	**5**	6	7	8	9	**10**
11	12	13	14	**15**	16	17	18	19	**20**
21	22	23	24	**25**	26	27	28	29	**30**
31	32	33	34	**35**	36	37	38	39	**40**
41	42	43	44	**45**	46	47	48	49	**50**
51	52	53	54	**55**	56	57	58	59	**60**
61	62	63	64	**65**	66	67	68	69	**70**
71	72	73	74	**75**	76	77	78	79	**80**
81	82	83	84	**85**	86	87	88	89	**90**
91	92	93	94	**95**	96	97	89	99	**100**

Multiples of 5 appear as 2 straight columns in the grid.

6 Times Tables

	1	2	3	4	5	6	7	8	9	10	11	12
6x	6	12	18	24	30	36	42	48	54	60	66	72

Multiples of 6 are always even numbers.

6 x Table

1	2	3	4	5	6	7	8	9	10
11	**12**	13	14	15	16	17	**18**	19	20
21	22	23	**24**	25	26	27	28	29	**30**
31	32	33	34	35	**36**	37	38	39	40
41	**42**	43	44	45	46	47	**48**	49	50
51	52	53	**54**	55	56	57	58	59	**60**
61	62	63	64	65	**66**	67	68	69	70
71	**72**	73	74	75	76	77	**78**	79	80
81	82	83	**84**	85	86	87	88	89	**90**
91	92	93	94	95	**96**	97	98	99	100

The last digit ends in 6 on the first row.
The last digit ends in 2 or 8 on the second row.
The last digit ends in 4 or 0 on the third row.

This pattern repeats.

6 Times Tables and Even Numbers

6 times an even number ends in that number.

6 x 2 = 12
6 x 4 = 24
6 x 6 = 36
6 x 8 = 48
6 x 10 = 60
6 x 12 = 72

6 Times Tables and Other Tables

	1	2	3	4	5	6	7	8	9	10	11	12
2x	2	4	6	8	10	12	14	16	18	20	22	24
3x	3	6	9	12	15	18	21	24	27	30	33	36
6x	6	12	18	24	30	36	42	48	54	60	66	72

The 6x table is twice the 3x table.
The 6x table is 3 times the 2x table.

7 Times Tables

	1	2	3	4	5	6	7	8	9	10	11	12
7x	7	14	21	28	35	42	49	56	63	70	77	84

Starting with the first, every second multiple of 7 is odd.

Starting with the second, every second multiple of 7 is even.

7 x Table										
1	2	3	4	5	6	**7**	8	9	10
11	12	13	**14**	15	16	17	18	19	20
21	22	23	24	25	26	27	**28**	29	30
31	32	33	34	**35**	36	37	38	39	40
41	**42**	43	44	45	46	47	48	**49**	50
51	52	53	54	55	**56**	57	58	59	60
61	62	**63**	64	65	66	67	68	69	**70**
71	72	73	74	75	76	**77**	78	79	80
81	82	83	**84**	85	86	87	88	89	90
91	92	93	94	95	96	97	**98**	99	100

Multiples of 7 appear as broken diagonal lines.

Tip

To help learn **7 x 8 = 56** remember the numbers 5,6,7,8.

8 Times Tables

	1	2	3	4	5	6	7	8	9	10	11	12
8x	8	16	24	32	40	48	56	64	72	80	88	96

Multiples of 8 are always even numbers.

8 x Table										
1	2	3	4	5	6	7	**8**	9	10
11	12	13	14	15	**16**	17	18	19	20
21	22	23	**24**	25	26	27	28	29	30
31	**32**	33	34	35	36	37	38	39	**40**
41	42	43	44	45	46	47	**48**	49	50
51	52	53	54	55	**56**	57	58	59	60
61	62	63	**64**	65	66	67	68	69	70
71	**72**	73	74	75	76	77	78	79	**80**
81	82	83	84	85	86	87	**88**	89	90
91	92	93	94	95	**96**	97	98	99	100

Multiples of 8 appear as broken diagonal lines.

There is one multiple of 8 per row, except for every fourth row.

- The last digit ends in 8 on the first row.
- The last digit ends in 6 on the second row.
- The last digit ends in 4 on the third row.

Times Tables Tricks and Tips

- The last digit ends in 2 or 0 on the fourth row.

This pattern repeats.

	1	2	3	4	5	6	7	8	9	10	11	12
2x	2	4	6	8	10	12	14	16	18	20	22	24
4x	4	8	12	16	20	24	28	32	36	40	44	48
8x	8	16	24	32	40	48	56	64	72	80	88	96

8 Times Tables and Other Tables

The 8x table is twice the 4x table.
The 8x table is 4 times the 2x table.

Tip

8 rows x 8 columns make a chessboard of 64 squares.

9 Times Tables

	1	2	3	4	5	6	7	8	9	10	11	12
9x	9	18	27	36	45	54	63	72	81	90	99	108

Starting with the first, every second multiple of 9 is odd.

Starting with the second, every second multiple of 9 is even.

The first digit increases by 1 each time.
The second digit decreases by 1 each time.

If you add the digits together, they always add up to 9.

This is a great help if you need to check your answer.

For example, use the multiple 27 which is 3 x 9.
2 + 7 = 9

Sometimes you need to add the digits more than once.

For example, use the multiple 99 which is 9 x 11.
9 + 9 = 18
We keep going until only one digit is left.
1 + 8 = 9

The final answer is 9!

There is an easy way to multiply by 9.

- Step 1: Multiply by 10 (very easy)
- Step 2: subtract the number you multiply by 9.

For example 4x9

- Step 1: **4 x 10 = 40**
- Step 2: **40 - 4 = 36**

9 x Table									
1	2	3	4	5	6	7	8	**9**	10
11	12	13	14	15	16	17	**18**	19	20
21	22	23	24	25	26	**27**	28	29	30
31	32	33	34	35	**36**	37	38	39	40
41	42	43	44	**45**	46	47	48	49	50
51	52	53	**54**	55	56	57	58	59	60
61	62	**63**	64	65	66	67	68	69	70
71	**72**	73	74	75	76	77	78	79	80
81	82	83	84	85	86	87	88	89	**90**
91	92	93	94	95	96	97	98	**99**	100
101	102	103	104	105	106	107	**108**	109	110

Multiples of 9 appear in a diagonal line in the grid.

9 Times Tables and Other Tables

	1	2	3	4	5	6	7	8	9	10	11	12
3x	3	6	9	12	15	18	21	24	27	30	33	36
6x	6	12	18	24	30	36	42	48	54	60	66	72
9x	9	18	27	36	45	54	63	72	81	90	99	108

The 9x table is one and a half times the 6x table.
The 9x table is 3 times the 3x table.

10 Times Tables

	1	2	3	4	5	6	7	8	9	10	11	12
10x	10	20	30	40	50	60	70	80	90	100	110	120

Multiples of 10 always end in zero.

10 Times Tables are easy. All we have to do to multiply by 10 is to place a zero at the end of the number.

For example, if we want to multiply 4 x 10, we put a zero after the 4 to get 40.

10 x Table									
1	2	3	4	5	6	7	8	9	10
11	12	13	14	15	16	17	18	19	20
21	22	23	24	25	26	27	28	29	30
31	32	33	34	35	36	37	38	39	40
41	42	43	44	45	46	47	48	49	50
51	52	53	54	55	56	57	58	59	60
61	62	63	64	65	66	67	68	69	70
71	72	73	74	75	76	77	78	79	80
81	82	83	84	85	86	87	88	89	90
91	92	93	94	95	96	97	89	99	100

Multiples of 10 appear in a single straight column in the grid.

Also the 10x tables are so easy we do not need to learn them. To multiply by 10 we just put a zero at the end of the number.

Multiplication Patterns

- If you multiply 2 even numbers you get an even number.
  For example, 2 x 4 = 8.

- If you multiply 2 odd numbers you get an odd number.
  For example, 3 x 7 = 21.

- If you multiply an even number by an odd number you get an even number.
  For example, 2 x 7 = 14.

11 Times Tables

	1	2	3	4	5	6	7	8	9	10	11	12
11x	11	22	33	44	55	66	77	88	99	110	121	132

Multiplying 11 by a single digit is easy. We repeat the digit.

For example, if we want to multiply 11 x 7, we repeat the number 7 to get 77.

To multiply 11 by a double digit number, we insert the sum of these two digits into the middle of the digits.

For example, to multiply 11 x 12, add the digits of 12 (1 and 2) to get 3.
Then insert 3 in the middle of 12 to get 132.

11 x Table									
1	2	3	4	5	6	7	8	9	10
11	12	13	14	15	16	17	18	19	20
21	**22**	23	24	25	26	27	28	29	30
31	32	**33**	34	35	36	37	38	39	40
41	42	43	**44**	45	46	47	48	49	50
51	52	53	54	**55**	56	57	58	59	60
61	62	63	64	65	**66**	67	68	69	70
71	72	73	74	75	76	**77**	78	79	80
81	82	83	84	85	86	87	**88**	89	90
91	92	93	94	95	96	97	98	**99**	100
101	102	103	104	105	106	107	108	109	**110**
111	112	113	114	115	116	117	118	119	120
121	122	123	124	125	126	127	128	129	130
131	**132**	133	134	135	136	137	138	139	140

Multiples of 11 appear as diagonal lines in the grid.

12 Times Tables

	1	2	3	4	5	6	7	8	9	10	11	12
12x	12	24	36	48	60	72	84	96	108	120	132	144

Multiples of 12 are always even numbers.

Times Tables Tricks and Tips

12 x Table									
1	2	3	4	5	6	7	8	9	10
11	**12**	13	14	15	16	17	18	19	20
21	22	23	**24**	25	26	27	28	29	30
31	32	33	34	35	**36**	37	38	39	40
41	42	43	44	45	46	47	**48**	49	50
51	52	53	54	55	56	57	58	59	**60**
61	62	63	64	65	66	67	68	69	70
71	**72**	73	74	75	76	77	78	79	80
81	82	83	**84**	85	86	87	88	89	90
91	92	93	94	95	**96**	97	98	99	100
101	102	103	104	105	106	107	**108**	109	110
111	112	113	114	115	116	117	118	119	**120**
121	122	123	124	125	126	127	128	129	130
131	**132**	133	134	135	136	137	138	139	140
141	142	143	**144**	145	146	147	148	149	150

A pattern here is that with each row you move down, the multiple moves 2 columns to the right.

Notice how this pattern is repeated twice. The first pattern is the numbers 12 to 60. The first pattern is the numbers 72 to 120. This pattern will continue forever.

12 Times Tables and Other Tables

The 12x table is twice the 6x table.
The 12x table is 3 times the 4x table.
The 12x table is 4 times the 3x table.
The 12x table is 6 times the 2x table.

Multiples of 12

Some multiples of 12 are used several times in the times table.

Multiples Table							
12	2x6	3x4	4x3	6x2			
24	2x12	3x8	4x6	6x4	8x3	12x2	
36	3x12	4x9	6x6	9x4	12x3		
48	3x16	4x12	6x8	8x6	9x4	12x4	16x3
64	4x16	8x8	16x4				

These are some of the numbers.

Cycles

Multiples of a number move in cycles. The last digit of a cycle repeats. The cycle restarts when a multiple ends in zero.

Lets look at the numbers in the 2x tables up to 2x10.

We get:
2,4,6,8,10,12,14,16,18,20

If we look at only the last digit we get:
2,4,6,8,0,2,4,6,8,0

We can see that after the digit is zero the same numbers repeat. This cycle goes for ever. For example the numbers up to 2x40 are:
2,4,6,8,10,12,14,16,18,20,22,24,26,28,30,32,34,36,38,40

Looking at the last digit we would get:
2,4,6,8,0,2,4,6,8,0,2,4,6,8,0,2,4,6,8,0, etc. (etc means and so on...forever)

The cycles in the Times Tables are:

Times Tables Tricks and Tips

- 2x table: 2,4,6,8,0
- 3x table: 3,6,9,2,5,8,1,4,7,0
- 4x table: 4,8,2,6,0
- 5x table: 5,0
- 6x table: 6,2,8,4,0
- 7x table: 7,4,1,8,5,2,9,3,0
- 8x table: 8,6,4,2,0
- 9x table: 9,8,7,6,5,4,3,2,1,0
- 10x table: 0

Tables which have a short cycle tend to be simpler. The 10x has the shortest possible cycle which is reason why it is so simple. The 5x table is also very short.

The even numbers (apart from 10, which is a 2 digit number in any case) each have a cycle of 5 numbers. The odd numbers (apart from 5) each have a cycle length of 10.

In the **9x** table, the last digit decreases by 1 each time. (10 - 9 = 1)

In the **8x** table, the last digit decreases by 2 each time. (10 - 8 = 2)

In the **2x** table, the last digit increases by 2 each time.

In the **7x** table, the last digit decreases by 3 each time. (10 - 7 = 3) If the result is less than zero, then 10 is added. For example the cycle goes from 1 to 8, 1 minus 3 is less than zero, so we add 10 to 1, then 11-3 = 8

In the **6x** table, the last digit decreases by 4 each time. (10 - 6 = 4) If the result is less than zero, then 10 is added. For example the cycle goes from 2 to 8, 2 minus 4 is less than zero, so we add 10 to 2, then 12-4 = 8

In the **3x** table, the last digit increases by 3 each time.

In the **4x** table, the last digit increases by 4 each time.

Adding digits together we get some new cycles: 6x 6,3,9,

Prime Numbers

1	2	3	4	5	6	7	8	9	10
11	12	13	14	15	16	17	18	19	20
21	22	23	24	25	26	27	28	29	30
31	32	33	34	35	36	37	38	39	40
41	42	43	44	45	46	47	48	49	50
51	52	53	54	55	56	57	58	59	60
61	62	63	64	65	66	67	68	69	70
71	72	73	74	75	76	77	78	79	80
81	82	83	84	85	86	87	88	89	90
91	92	93	94	95	96	97	98	99	100

Some numbers are never used in the Times Tables. One example is **prime** numbers.

Prime numbers are numbers which cannot be divided by any numbers, besides 1 and itself.

2 is the only even prime number. All other prime numbers are odd.

You do not memorise the prime numbers. However, its good to know some numbers are not answers in the times tables.

Squares

Square numbers are numbers which is the result of multiplying a number by itself. They can used to form a square shape.

Understanding the square numbers can help you learn the times tables.

Times Tables Tricks and Tips

s

The square of **2** is **4**.

The square of **3** is **9**.

1	2	3	4
5	6	7	8
9	10	11	12
13	14	15	**16**

The square of **4** is **16**.

1	2	3	4	5
6	7	8	9	10
11	12	13	14	15
16	17	18	19	20
21	22	23	24	**25**

The square of **5** is **25**.

Times Tables Tricks and Tips

1	2	3	4	5	6
7	8	9	10	11	12
13	14	15	16	17	18
19	20	21	22	23	24
25	26	27	28	29	30
31	32	33	34	35	**36**

The square of **6** is **36**.

1	2	3	4	5	6	7
8	9	10	11	12	13	14
15	16	17	18	19	20	21
22	23	24	25	26	27	28
29	30	31	32	33	34	35
36	37	38	39	40	41	42
43	44	45	46	47	48	**49**

The square of **7** is **49**.

1	2	3	4	5	6	7	8
9	10	11	12	13	14	15	16
17	18	19	20	21	22	23	24
25	26	27	28	29	30	31	32
33	34	35	36	37	38	39	40
41	42	43	44	45	46	47	48
49	50	51	52	53	54	55	56
57	58	59	60	61	62	63	**64**

The square of **8** is **64**.

1	2	3	4	5	6	7	8	9
10	11	12	13	14	15	16	17	18
19	20	21	22	23	24	25	26	27
28	29	30	31	32	33	34	35	36
37	38	39	40	41	42	43	44	45
46	47	48	49	50	51	52	53	54
55	56	57	58	59	60	61	62	63
64	65	66	67	68	69	70	71	72
73	74	75	76	77	78	79	80	**81**

The square of **9** is **81**.

Times Tables Tricks and Tips

1	2	3	4	5	6	7	8	9	10
11	12	13	14	15	16	17	18	19	20
21	22	23	24	25	26	27	28	29	30
31	32	33	34	35	36	37	38	39	40
41	42	43	44	45	46	47	48	49	50
51	52	53	54	55	56	57	58	59	60
61	62	63	64	65	66	67	68	69	70
71	72	73	74	75	76	77	78	79	80
81	82	83	84	85	86	87	88	89	90
91	92	93	94	95	96	97	98	99	**100**

The square of **10** is **100**.

1	2	3	4	5	6	7	8	9	10	11
12	13	14	15	16	17	18	19	20	21	22
23	24	25	26	27	28	29	30	31	32	33
34	35	36	37	38	39	40	41	42	43	44
45	46	47	48	49	50	51	52	53	54	55
56	57	58	59	60	61	62	63	64	65	66
67	68	69	70	71	72	73	74	75	76	77
78	79	80	81	82	83	84	85	86	87	88
89	90	91	92	93	94	95	96	97	98	99
100	101	102	103	104	105	106	107	108	109	110
111	112	113	114	115	116	117	118	119	120	**121**

The square of **11** is **121**.

1	2	3	4	5	6	7	8	9	10	11	12
13	14	15	16	17	18	19	20	21	22	23	24
25	26	27	28	29	30	31	32	33	34	35	36
37	38	39	40	41	42	43	44	45	46	47	48
49	50	51	52	53	54	55	56	57	58	59	60
61	62	63	64	65	66	67	68	69	70	71	72
73	74	75	76	77	78	79	80	81	82	83	84
85	86	87	88	89	90	91	92	93	94	95	96
97	98	99	100	101	102	103	104	105	106	107	108
109	110	111	112	113	114	115	116	117	118	119	120
121	122	123	124	125	126	127	128	129	130	131	132
133	134	135	136	137	138	139	140	141	142	143	144

The square of **12** is **144**.

	2	3	4	5	6	7	8	9	10	11	12
2	4	6	8	10	12	14	16	18	20	22	24
3	6	9	12	15	18	21	24	27	30	33	36
4	8	12	16	20	24	8	32	36	40	44	48
5	10	15	20	25	30	35	40	45	50	55	60
6	12	18	24	30	36	42	48	54	60	66	72
7	14	21	28	35	42	49	56	63	70	77	84
8	16	24	32	40	48	56	64	72	80	88	96
9	18	27	36	45	54	63	72	81	90	99	108
10	20	30	40	50	60	70	80	90	100	110	120
11	22	33	44	55	55	77	88	99	110	121	132
12	24	36	48	60	72	84	96	108	121	132	144

Times Tables Tricks and Tips

Square numbers are highlighted. They form a diagonal line from the top left corner to the bottom right corner.

1 4 9 16 25 36 49 64 81 100 121 144

Here are the squares of the numbers from 1 to 12.

The difference between the squares of 1 and 2 is 3. (4 - 1 = 3)

The differences between the squares look like this:

3 5 7 9 11 13 15 17 19

If we put zero at the front of list of squares we get

1 3 5 7 9 11 13 15 17 19

This is a sequence of odd numbers.

1 2 3
4 5 6
7 8 9

Here we have numbers representing the square of 3.

To make a square one number larger, we add one row and one column.
This will add 3 new squares for the row which will placed under the 7, 8 and 9.
There will be under 3 new squares added to the right of 3, 6 and 9. So we are adding twice the old number (3), plus one for the square in the bottom right hand corner (where the x is).

If we know the square of one number we can work out the square of the next number. We can do this by adding twice the first number plus one.

1	2	3	4	5	
6	7	8	9	10	
11	12	13	14	15	
16	17	18	19	20	
21	22	23	24	25	

For example, if we know the square of 5 is 25, we can get the square of 6.
It is 25 + 2x5 + 1 = 36

A more complicated example:

We know the square of 1,000 is 1,000,000.
Therefore we know the square of 1,001 is 1,000,000 + 2,000 + 1 = 1,002,001.

Division

What is Division?

The multiplication table can also be used for division, though in a different way.
This works best if the result is a whole number.

Times Tables Tricks and Tips

	1	2	3	4	5	6	7	8	9	10	11	12
1x	1	2	3	4	5	6	7	8	9	10	11	12
2x	2	4	6	8	10	12	14	16	18	20	22	24
3x	3	6	9	12	15	18	21	24	27	30	33	36
4x	4	4	8	16	20	24	28	32	36	40	44	48
5x	5	10	15	20	25	30	35	40	45	50	55	60
6x	6	12	18	24	30	36	42	48	54	60	66	72
7x	7	14	21	28	35	42	49	56	63	70	77	84
8x	8	16	24	32	40	48	56	64	72	80	88	96
9x	9	18	27	36	45	54	63	72	81	90	99	108
10x	10	20	30	40	50	60	70	80	90	100	110	120
11x	11	22	33	44	55	66	77	88	99	110	121	132
12x	12	24	36	48	60	72	84	96	108	120	132	144

- Find the row of the number you want to divide by.
- Search in the row for the number you want to divide.
- Find the column of that number (if it exists). That is the answer.

Say you want to divide 56 by 8.
Find the **8x** row.
Find 56 in that row.
The column of that row is 7, which is the answer.

Division with a Remainder

Say you want to divide 50 by 8. 8 does not go exactly, there will be a **remainder**.

You can use your knowledge of the 8x tables to find the closest number 8 goes into.
8x6=48 and 8x7=56.
48 is lower than 50, so the answer is 6 plus a remainder.

The remainder is the **difference** between 50 and 48.
50-48=2 so 2 is the remainder.
The answer is 6 remainder 2.

Division by 10

Division by 10. Remove the last zero.

When a number whose last digit is not 0, is divided by 10, the result is a fraction.

Division by 5

Double the number. Divide by 10.

For example: What is 45 divided by 5?
Doubling 45 is 45 x 2 = 90.
Dividing 90 by 10 is 90/10 = 9.

When a number whose last digit is not 0 or 5, is divided by 5, the result is a fraction.

Square Roots

The square root of a number is the opposite of the square of a number. If we know the square of 2 is 4, we know that the square root of 4 is 2.

The square roots of squares are easy:

Square	1	4	9	16	25	36	49	64	81	100	121	144
Square Root	1	2	3	4	5	6	7	8	9	10	11	12

The square root of a number in the first row is in the second row. For example, the square root of 49 is 7.

Square roots of numbers which are not perfect squares are fractions and more difficult to work out.

More Plus Tables

These plus tables add up to 11 or more.

11 Plus Table

	1	2	3	4	5	6	7	8	9	10	11
2+9=11	2								9		
3+8=11			3					8			
4+7=11				4			7				
5+6=11					5	6					
6+5=11						6	5				
7+4=11							7	4			
8+3=11								8	3		
9+2=11									9		2

12 Plus Table

	1	2	3	4	5	6	7	8	9	10	11	12
3+9=12			3						9			
4+8=12				4				8				
5+7=12					5		7					
6+6=12						6	6					
7+5=12							7	5				
8+4=12								8	4			
9+3=12									9			3

13 Plus Table

	1	2	3	4	5	6	7	8	9	10	11	12	13
4+9=13				4					9				
5+8=13					5				8				
6+7=13						6		7					
7+6=13							7			6			
8+5=13								8			5		
9+4=13									9				4

14 Plus Table

	1	2	3	4	5	6	7	8	9	10	11	12	13	14
5+9=14					5				9					
6+8=14						6		8						
7+7=14							7		7					
8+6=14								8		6				
9+5=14									9			5		

15 Plus Table

	1	2	3	4	5	6	7	8	9	10	11	12	13	14	15
6+9=15						6			9						
7+8=15							7	8							
8+7=15								8		7					
9+6=15									9			6			

16 Plus Table

	1	2	3	4	5	6	7	8	9	10	11	12	13	14	15	16
8+8=16					8							8				
9+7=16					9							7				

The Minus Tables

2 Minus Table

	1	2
2-1=1		

3 Minus Table

	1	2	3
3-1=2		2	1
3-2=1	1	2	

4 Minus Table

	1	2	3	4
4-1=3		3		1
4-2=2	2		2	
4-3=1	1	3		

5 Minus Table

	1	2	3	4	5
5-1=4	4				1
5-2=3	3		2
5-3=2	2	3
5-4=1	1	4

6 Minus Table

	1	2	3	4	5	6
6-1=5	5			1
6-2=4	4		2
6-3=3	3	3
6-4=2	2	4
6-5=1	1	5

7 Minus Table

	1	2	3	4	5	6	7
7-1=6	6		1
7-2=5	5	2
7-3=4	4	3
7-4=3	3	4
7-5=2	2	5
7-6=1	1	6

Times Tables Tricks and Tips

8 Minus Table								
	1	2	3	4	5	6	7	8
8-1=7	7							1
8-2=6	6						2	
8-3=5	5					3		
8-4=4	4				4			
8-5=3	3			5				
8-6=2	2		6					
8-7=1	1	7						

9 Minus Table									
	1	2	3	4	5	6	7	8	9
9-1=8	8								1
9-2=7	7							2	
9-3=6	6						3		
9-4=5	5					4			
9-5=4	4				5				
9-6=3	3			6					
9-7=2	2		7						
9-8=1	1	8							

FM Bill Jordan

10 Minus Table

	1	2	3	4	5	6	7	8	9	10
10-1=9	9									1
10-2=8	8								2	
10-3=7	7							3		
10-4=6	6						4			
10-5=5	5					5				
10-6=4	4				6					
10-7=3	3			7						
10-8=2	2		8							
10-9=1	1	9								

Times Tables Tricks and Tips

More Minus Tables

	11 Minus Table									
	1	2	3	4	5	6	7	8	9	10	11
11-2=9		9									2
11-3=8		8						3		
11-4=7		7					4			
11-5=6		6				5				
11-6=5		5			6					
11-7=4		4		7						
11-8=3		3	8							
11-9=2		2	9								

12 Minus Table

	1	2	3	4	5	6	7	8	9	10	11	12
12-3=9					9					3		
12-4=8				8					4			
12-5=7				7				5				
12-6=6			6				6					
12-7=5		5				7						
12-8=4		4			8							
12-9=3	3			9								

13 Minus Table

	1	2	3	4	5	6	7	8	9	10	11	12	13
13-4=9					9					4			
13-5=8				8					5				
13-6=7			7				6						
13-7=6		6				7							
13-8=5		5			8								
13-9=4	4			9									

Times Tables Tricks and Tips

14 Minus Table

	1	2	3	4	5	6	7	8	9	10	11	12	13	14
14-5=9					9						5			
14-6=8					8							6		
14-7=7					7								7	
14-8=6				6									8	
14-9=5				5										9

15 Minus Table

	1	2	3	4	5	6	7	8	9	10	11	12	13	14	15
15-6=9						9						6			
15-7=8						8							7		
15-8=7					7									8	
15-9=6				6											9

16 Minus Table

	1	2	3	4	5	6	7	8	9	10	11	12	13	14	15	16
16-7=9							9						7			
16-8=8								8						8		
16-9=7									7						9	

17 Minus Table

	1	2	3	4	5	6	7	8	9	10	11	12	13	14	15	16	17
17-8=9								9						8			
17-9=8									8						9		

18 Minus Table

	1	2	3	4	5	6	7	8	9	10	11	12	13	14	15	16	17	18
18-9=9									9						9			

Summary

- Introduction
- The Grid
- Odds and Evens
- The plus Tables
- The Minus Tables
- What is Multiplication?
- How many Times Tables are there?
- Multiplying by One
- Multiplying by Ten
- How many items to learn?
- 2 Times Tables
- 3 Times Tables
- 4 Times Tables
- 5 Times Tables
- 6 Times Tables
- 7 Times Tables
- 8 Times Tables
- 9 Times Tables
- 10 Times Tables
- 11 Times Tables
- 12 Times Tables
- Cycles
- Prime Numbers
- Squares
- Division
- Square Roots
- More Plus Tables
  2 digits adding up to 11 or more.
- More Minus Tables

Where to go from here?

I hope you have enjoyed my book. It includes some old approaches and a few experimental ones. It is designed to read more than once. If you have any questions or suggestions you can email Bill Jordan at **swneerava@gmail.com**. Small changes are easy to implement.

About the Author

Bill Jordan is a self-taught programmer who created a large number of programs for the Public Domain. He has worked as a programmer for several companies and on various programs for others. He was an IT trainer for 10 years tutoring adults.

He is also a national senior chess champion and has written a number of chess books and other books available on Amazon.

www.ingramcontent.com/pod-product-compliance
Lightning Source LLC
Chambersburg PA
CBHW070849220526
45466CB00005B/1938